Math Mammoth
Grade 2 Tests and
Cumulative Reviews

for the complete curriculum
(Canadian Version)

Includes consumable student copies of:

- Chapter Tests
- End-of-year Test
- Cumulative Reviews

By Maria Miller

ISBN 978-1-954358-21-8

Contents

Grade 2, Chapter 1

End-of-Chapter Test

Instructions to the student:

Answer each question in the space provided.

Instructions to the teacher:

My suggestion for grading the chapter 1 test is below. The total is 24 points. Divide the student's score by the total of 24 to get a decimal number, and change that decimal to percent to get the student's percentage score.

Question #	Max. points	Student score
1	8 points	
2	2 points	
3	6 points	

Question #	Max. points	Student score
4	2 points	
5	6 points	
Total	24 points	

Chapter 1 Test

1. Add and subtract.

a.	b.	c.	d.
$58 + 2 =$ _____	$33 + 4 =$ _____	$50 + 6 =$ _____	$45 + 40 =$ _____
$58 - 6 =$ _____	$94 - 4 =$ _____	$65 - 30 =$ _____	$98 - 70 =$ _____

2. Colour.

a. The fourth flower from the left.	b. The sixth flower from the right.

3. Fill in the missing numbers. The four problems form a fact family.

a. $2 + $ ____ $= 10$

   ____ $+ $ ____ $= 10$

   $10 - $ ____ $=$ ____

   ____ $- $ ____ $=$ ____

b. ____ $+ $ ____ $= 9$

   ____ $+ $ ____ $= 9$

   ____ $- $ ____ $=$ ____

   $9 - $ ____ $= 6$

c. ____ $+ $ ____ $=$ ____

   ____ $+ $ ____ $=$ ____

   $15 - 7 = 8$

   ____ $- $ ____ $=$ ____

4. You read 23 pages in a story book. Your friend Susan read double that many pages. How many pages did Susan read?

5. Are these numbers even or odd? Mark an "X".

Number	Even?	Odd?
4		
10		

Number	Even?	Odd?
9		
16		

Number	Even?	Odd?
11		
18		

Grade 2, Chapter 2

End-of-Chapter Test

Instructions to the student:

Answer each question in the space provided.

Instructions to the teacher:

My suggestion for grading the chapter 2 test is below. The total is 29 points. Divide the student's score by the total of 29 to get a decimal number, and change that decimal to percent to get the student's percentage score.

Question #	Max. points	Student score
1	16 points	
2	4 points	
3	4 points	

Question #	Max. points	Student score
4	5 points	
Total	29 points	

Chapter 2 Test

1. Write the time with hours:minutes, and using "past", "to", "half past" or "o'clock".

a.	b.	c.	d.
_____ : _____	_____ : _____	_____ : _____	_____ : _____
_____ past _____	_____	_____	_____
e.	f.	g.	h.
_____ : _____	_____ : _____	_____ : _____	_____ : _____
_____ till _____	_____	_____	_____

2. Write the later time.

Time now	3:50	7:25
5 minutes later		

Time now	9 AM	12 noon
1 hour later		

3. Write the time using the **hours:minutes** way.

a. 20 past 4	**b.** 15 past 11	**c.** 15 to 12	**d.** 25 to 7
_____ : _____	_____ : _____	_____ : _____	_____ : _____

4. How many hours pass?

from	5 AM	8 AM	2 AM	10 AM	11 AM
to	12 noon	2 PM	3 PM	10 PM	6 PM
hours					

Grade 2, Chapter 3

End-of-Chapter Test

Instructions to the student:

Answer each question in the space provided.

Instructions to the teacher:

My suggestion for grading the chapter 3 test is below. The total is 31 points. Divide the student's score by the total of 31 to get a decimal number, and change that decimal to percent to get the student's percentage score.

Question #	Max. points	Student score
1	12 points	
2	6 points	
3	3 points	

Question #	Max. points	Student score
4	6 points	
5	4 points	
Total	31 points	

Chapter 3 Test

1. Add and find the missing numbers.

a. $9 + 6 =$ _____ $9 + 4 =$ _____	**b.** $8 + 9 =$ _____ $7 + 5 =$ _____	**c.** $7 +$ _____ $= 14$ $7 +$ _____ $= 16$
d. $9 +$ _____ $= 12$ $9 +$ _____ $= 18$	**e.** $8 + 5 =$ _____ $6 + 7 =$ _____	**f.** $6 + 8 =$ _____ $8 + 7 =$ _____

2. Subtract. For each problem write a corresponding addition fact.

a. $14 - 5 =$ _____ ____ $+$ ____ $=$ ____	**b.** $11 - 8 =$ _____ ____ $+$ ____ $=$ ____	**c.** $17 - 8 =$ _____ ____ $+$ ____ $=$ ____

3. Write $<$, $>$, or $=$.

    a. $7 + 9$ ☐ $8 + 8$ **b.** $40 - 5$ ☐ $40 - 8$ **c.** $\frac{1}{2}$ of 20 ☐ $\frac{1}{2}$ of 18

4. Subtract.

a. $11 - 6 =$ _____ $17 - 9 =$ _____	**b.** $16 - 8 =$ _____ $14 - 8 =$ _____	**c.** $13 - 6 =$ _____ $15 - 8 =$ _____

5. Solve.

a. Angela has 7 more teddy bears than Jerry. Jerry has 9.
How many does Angela have?

b. You have saved \$7 towards buying a toy that costs \$14.
Then, Grandmother gives you \$5.
How much money do you still need?

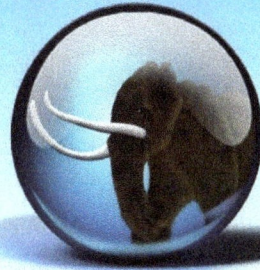

Grade 2, Chapter 4

End-of-Chapter Test

Instructions to the student:

Answer each question in the space provided.

Instructions to the teacher:

My suggestion for grading the chapter 4 test is below. The total is 21 points. Divide the student's score by the total of 21 to get a decimal number, and change that decimal to percent to get the student's percentage score.

Question #	Max. points	Student score
1	5 points	
2	6 points	
3	4 points	

Question #	Max. points	Student score
4	6 points	
Total	21 points	

Chapter 4 Test

1. Add.

a.
$$\begin{array}{r} 3\ 9 \\ +\ 4\ 6 \\ \hline \end{array}$$

b.
$$\begin{array}{r} 8\ 3 \\ 1\ 4 \\ +\ 2\ 5 \\ \hline \end{array}$$

c.
$$\begin{array}{r} 4\ 6 \\ 8 \\ +\ 3\ 3 \\ \hline \end{array}$$

d.
$$\begin{array}{r} 3\ 8 \\ +\ 2\ 3 \\ \hline \end{array}$$

e.
$$\begin{array}{r} 2\ 4 \\ 7 \\ 5\ 8 \\ +\ 1\ 5 \\ \hline \end{array}$$

2. Add.

a. $52 + 7 = $ _____

$18 + 5 = $ _____

b. $67 + 6 = $ _____

$27 + 8 = $ _____

c. $88 + 5 = $ _____

$43 + 8 = $ _____

3. Add mentally.

a. $2 + 6 + 8 + 7 = $ _____

$5 + 7 + 4 + 8 = $ _____

b. $42 + 2 + 10 + 5 = $ _____

$30 + 30 + 9 + 7 = $ _____

4. Solve the problems.

a. Mary worked for 28 hours and Brenda worked for 13. How many more hours did Mary work than Brenda?

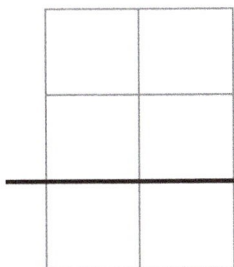

b. Find the total cost if you buy a stuffed animal for $12 and two books for $17 each.

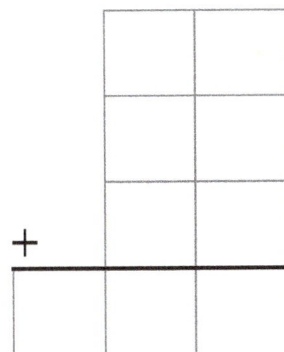

c. Fifteen birds were perched on a tree. Then, nine more birds flew in, and two birds flew away. How many birds are in the tree now?

Grade 2, Chapter 5

End-of-Chapter Test

Instructions to the student:

Answer each question in the space provided.

Instructions to the teacher:

My suggestion for grading the chapter 5 test is below. The total is 22 points. Divide the student's score by the total of 22 to get a decimal number, and change that decimal to percent to get the student's percentage score.

Question #	Max. points	Student score
1	4 points	
2	2 points	
3	2 points	
4	4 points	

Question #	Max. points	Student score
5	4 points	
6	6 points	
Total	22 points	

Chapter 5 Test

1. Join the dots with lines. Use a ruler. Write the name of the shape you get.

a. _____

b. _____

c. _____

d. _____

2. Draw a rectangle or a square so it encloses the given number of smaller squares.

a.

16 little squares

b.

9 little squares

3. Design a pattern with rectangles and/or squares.

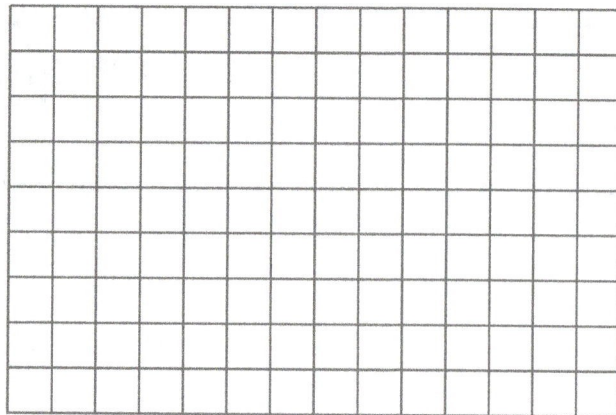

4. Write the fraction.

 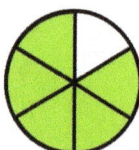

a. ——   b. ——   c. ——   d. ——

5. Divide these shapes. Then colour as you are asked to.

a.	b.	c.	d.
Divide this into thirds. Colour $\frac{1}{3}$.	Divide this into halves. Colour $\frac{2}{2}$.	Divide this into halves. Colour $\frac{1}{2}$.	Divide this into fourths. Colour $\frac{3}{4}$.

6. Colour. Then compare and write $<$, $>$, or $=$.

a. $\frac{1}{3}$ $\frac{2}{5}$ b. $\frac{5}{6}$ $\frac{3}{4}$ c. $\frac{3}{3}$ $\frac{4}{4}$

Grade 2, Chapter 6

End-of-Chapter Test

Instructions to the student:

Answer each question in the space provided.

Instructions to the teacher:

My suggestion for grading the chapter 6 test is below. The total is 33 points. Divide the student's score by the total of 33 to get a decimal number, and change that decimal to percent to get the student's percentage score.

Question #	Max. points	Student score
1	5 points	
2	2 points	
3	4 points	
4	4 points	
5	2 points	

Question #	Max. points	Student score
6	4 points	
7	3 points	
8	6 points	
9	3 points	
Total	33 points	

Chapter 6 Test

1. **a.** Count by fives:

475, 480, _____, _____, _____, _____, _____

   b. Count by tens:

376, 386, _____, _____, _____, _____, _____

2. Break these numbers into hundreds, tens, and ones.

a.	**b.**
235 = _____ + _____ + _____	805 = _____ + _____ + _____

3. These numbers are broken into their hundreds, tens, and ones. Write the numbers.

a. $600 + 80 + 8 =$ _____	**b.** $80 + 200 + 5 =$ _____
$400 + 60 =$ _____	$100 + 6 =$ _____

4. Write either $<$ or $>$ between the numbers.

a. 159 300	**b.** 323 230	**c.** 450 504	**d.** 482 284

5. Arrange the numbers in order.

a. 689, 869, 986	**b.** 524, 245, 452
_____ < _____ < _____	_____ < _____ < _____

6. Compare the expressions and write $<$, $>$, or $=$.

   a. $6 + 200 + 50$ ☐ 256 **b.** $800 + 9$ ☐ $90 + 800$

   c. $400 + 60 + 2$ ☐ $40 + 6 + 200$ **d.** $3 + 700$ ☐ $700 + 6$

7. One of the "parts" for the numbers is missing. Find out what number the triangle means.

a. 300 + △ + 7 = 347 △ = _____	b. 900 + △ + 40 = 948 △ = _____	c. 5 + △ + 80 = 585 △ = _____

8. Add and subtract.

a.	b.	c.
765 − 200 = _____	802 − 400 = _____	778 − 500 = _____
548 − 300 = _____	980 − 600 = _____	994 − 900 = _____

9. Some children counted cars that were passing by while waiting for the bus.

One in the pictograph means 5 cars.

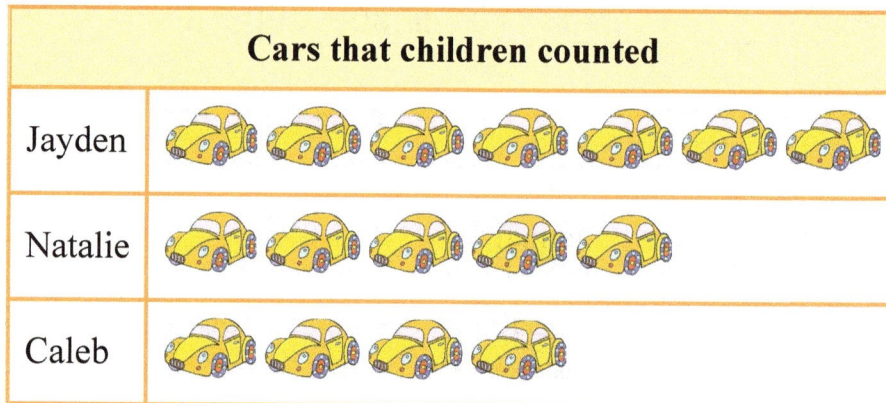

Cars that children counted

Jayden

Natalie

Caleb

 = 5 cars

a. How many cars did Natalie count?

b. How many did Jayden count?

c. How many more did Natalie count than Caleb?

Grade 2, Chapter 7

End-of-Chapter Test

Instructions to the student:

Answer each question in the space provided.

Instructions to the teacher:

My suggestion for grading the chapter 7 test is below. The total is 17 points. Divide the student's score by the total of 17 to get a decimal number, and change that decimal to percent to get the student's percentage score.

Question #	Max. points	Student score
1	4 points	
2	2 points	
3	2 points	

Question #	Max. points	Student score
4	1 point	
5	4 points	
6	4 points	
Total	17 points	

Chapter 7 Test

1. Cross out the sentences that don't make sense.

   a. An 11-year old boy weighs 12 kg.   **b.** An elephant is 3 m tall.

   c. My science book is 25 m wide.   **d.** The suitcase weighs 400 kg.

2. Measure these pencils to the nearest centimetre.

   #1

   #2

Pencil #1	cm
Pencil #2	cm

3. **a.** Draw a line that is 11 cm long.

   b. Draw a line that is 9 cm long.

4. Arrange these measuring units from the shortest to the longest.

   kilometre   centimetre   metre

5. Which unit do we use to measure these: centimetres (cm), metres (m), or kilometres (km)?

Distance	Unit	Distance	Unit
from Hong Kong to Rome		length of a garden	
around your head		height of a room	

6. The teacher needs to arrange this task beforehand, and check student's results.

   Ask your teacher to hand you an item. How much does it weigh? _____

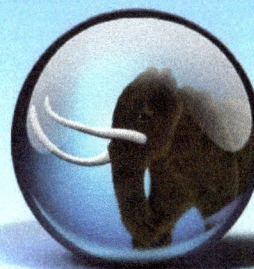

Grade 2, Chapter 8

End-of-Chapter Test

Instructions to the student:

Answer each question in the space provided.

Instructions to the teacher:

My suggestion for grading the chapter 8 test is below. The total is 27 points. Divide the student's score by the total of 27 to get a decimal number, and change that decimal to percent to get the student's percentage score.

Question #	Max. points	Student score
1	4 points	
2	4 points	
3	6 points	

Question #	Max. points	Student score
4	2 points	
5	11 points	
Total	27 points	

Chapter 8 Test

1. Add and subtract.

a.   2 1 9
    +4 3 5
	b. 5 6 2
	    +3 7 5
		c. 4 9 6
		    +2 8 6
			d. 6 2
			    − 2 7

2. Subtract. Check by adding the result and what was subtracted.

a.   9 6 4
   −2 2 7   + _____
	b. 7 4 8
	   −3 7 2 + _____

3. Add and subtract mentally.

a.	b.	c.
80 + 40 = _____	690 + 60 = _____	93 − 52 = _____
280 + 50 = _____	85 − 31 = _____	91 − 89 = _____

4. Find the total that Natalie paid for three
   paintings that cost $120 each.

5. Solve the word problems.

a. In a storehouse there were 250 sacks of wheat.
Then the store owner sold 68 sacks.

How many sacks are left?

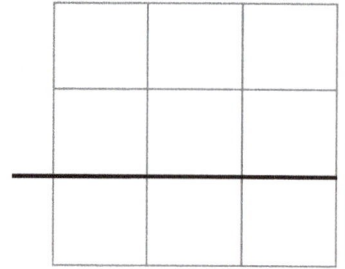

b. A pet store has 52 kittens. Of them, 15 are white
and 18 are orange. The rest are black.

How many are black?

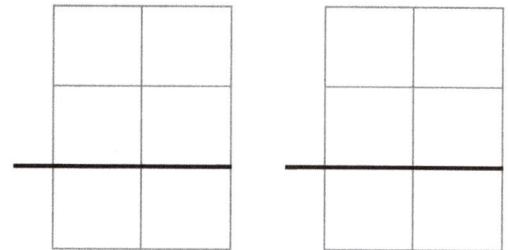

c. Regina bought two 8-kg bags of cat food, and one 25-kg bag of dog food.
How much is the total weight of these bags?

d. A store sold 47 coffee makers during the the month of January. During February,
they sold 19 fewer coffee makers.
How many coffee makers did the store
sell in February?

How many did they sell in those two months?

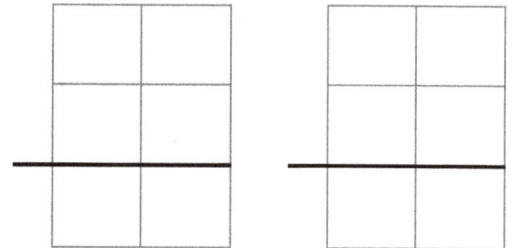

e. Grandfather walked 300 metres on Tuesday.
The next day he walked 120 metres more
than on Tuesday. How many metres did
he walk in those two days in total?

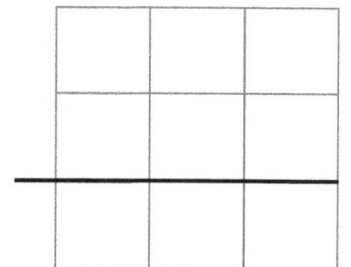

Grade 2, Chapter 9

End-of-Chapter Test

Instructions to the student:

Answer each question in the space provided.

Instructions to the teacher:

My suggestion for grading the chapter 9 test is below. The total is 10 points. Divide the student's score by the total of 10 to get a decimal number, and change that decimal to percent to get the student's percentage score.

Question #	Max. points	Student score
1	4 points	
2	2 points	

Question #	Max. points	Student score
4	4 points	
Total	10 points	

Chapter 9 Test

1. How much money? Write the amount.

a. $_____

b. $_____

c. $_____

d. $_____

2. Find the change.

a. $1.70

Customer gives $2.00

Change $_____

b. $2.85

Customer gives $5

Change $_____

3. Find the total cost.

a. Matt bought two sandwiches for $1.55 each and water for $0.75.

$
+
$

b. Eva bought two sets of water paints for $2.55 each.

$
+
$

Grade 2, Chapter 10

End-of-Chapter Test

Instructions to the student:

Answer each question in the space provided.

Instructions to the teacher:

My suggestion for grading the chapter 10 test is below. The total is 26 points. Divide the student's score by the total of 26 to get a decimal number, and change that decimal to percent to get the student's percentage score.

Question #	Max. points	Student score
1	6 points	
2	2 points	
3	2 points	

Question #	Max. points	Student score
4	4 points	
5	12 points	
Total	26 points	

Chapter 10 Test

1. Draw groups to illustrate the multiplication.

a. $6 \times 1 =$ _____	**b.** $2 \times 7 =$ _____	**c.** $3 \times 3 =$ _____

2. Write each addition as a multiplication.

    a. $6 + 6 + 6 + 6 =$ _____ $\times$ _____

    b. $50 + 50 + 50 =$ _____ $\times$ _____

3. Write each multiplication as an addition.

    a. $2 \times 8 =$ _____

    b. $5 \times 3 =$ _____

4. Draw number-line jumps for these multiplications.

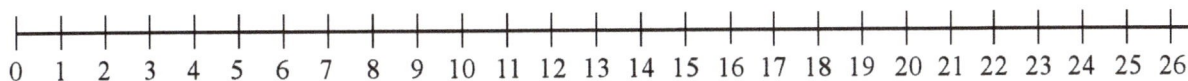

    a. $6 \times 3 =$ _____

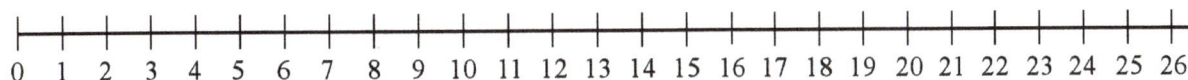

    b. $5 \times 5 =$ _____

5. Multiply.

a. $4 \times 3 =$ _____
 	
   $3 \times 10 =$ _____
	b. $5 \times 0 =$ _____
	
	   $2 \times 20 =$ _____
		c. $1 \times 6 =$ _____
		
		   $2 \times 9 =$ _____

d. $4 \times 3 =$ _____
 	
   $2 \times 8 =$ _____
	e. $2 \times 12 =$ _____
	
	   $3 \times 5 =$ _____
		f. $4 \times 10 =$ _____
		
		   $1 \times 800 =$ _____

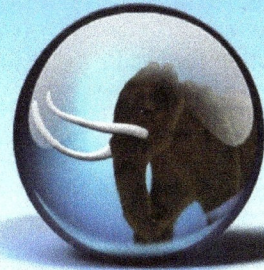

Math Mammoth End-of-Year Test - Grade 2
International Version (Canada)

This test is quite long, so I do not recommend having the student do it in one sitting. Break it into parts and administer them either on consecutive days, or perhaps in the morning/evening/morning. Use your judgement.

This is to be used as a diagnostic test. Thus, you may even skip those areas and concepts that you already know for sure your student has mastered.

The test checks for all major concepts covered in *Math Mammoth Grade 2*. This test is evaluating the student's ability in the following content areas:

- basic addition and subtraction facts within 0-18
- three-digit numbers and place value
- regrouping in addition with two- and three-digit numbers
- regrouping in subtraction with two- and three-digit numbers, excluding regrouping two times
- addition and subtraction
- basic word problems
- measuring and drawing with a ruler, to the nearest centimetre
- names and usage of units for measuring length and weight
- names of basic shapes
- the concept of a fraction
- reading the clock to the nearest five minutes
- counting coins and banknotes
- the concept of multiplication

Note: Problems #1 and #2 are done <u>orally and timed</u>. Let the student see the problems. Read each problem aloud, and wait a maximum of 5 seconds for an answer. Mark the problem as right or wrong according to the student's (oral) answer. Mark it wrong if there is no answer. Then you can move on to the next problem.

You do not have to mention to the student that the problems are timed or that he/she will have 5 seconds per answer, because the idea here is not to create extra pressure by the fact it is timed, but simply to check if the student has the facts memorized (quick recall). You can say for example (vary as needed):

"I will ask you some addition and subtraction questions. Try to answer them as quickly as possible. In each question, I will only wait a little bit for you to answer, and if you do not say anything, I will move on to the next problem. So just try your best to answer the questions as quickly as you can."

In order to continue with *Math Mammoth Grade 3*, I recommend that the student score at least 80% on this test, and that the teacher or parent revise with him any content areas in which he is weak. Students scoring between 70% and 80% may also continue with grade 3, depending on the types of errors (careless errors or not remembering something, versus lack of understanding). The most important areas to master are topics related to addition and subtraction, word problems, and place value. Again, use your judgement.

Grading

My suggestion for grading is below. The total is 134 points. A score of 107 points is 80%.

Question	Max. points	Student score
Basic Addition and Subtraction Facts		
1	16 points	
2	16 points	
3	6 points	
	subtotal	/ 38
Mental Addition and Subtraction with Two-Digit Numbers and Word Problems		
4	1 point	
5	2 points	
6	3 points	
7	1 point	
8	3 points	
9	3 points	
10	6 points	
	subtotal	/ 19
Three-Digit Numbers		
11	2 points	
12	2 points	
13	2 points	
14	6 points	
15	4 points	
	subtotal	/ 16
Regrouping in Addition and Subtraction, including Word Problems		
16	3 points	
17	4 points	
18	2 points	
19	2 points	
20	2 points	
21	3 points	
	subtotal	/ 16

Question	Max. points	Student score
Clock		
22	6 points	
23	5 points	
	subtotal	/ 11
Money		
24	2 points	
25	2 points	
26	2 points	
	subtotal	/ 6
Geometry and Measuring		
27	2 points	
28	4 points	
29	1 point	
30	4 points	
	subtotal	/ 11
Fractions		
31	4 points	
32	6 points	
	subtotal	/ 10
Concept of Multiplication		
33	2 points	
34	2 points	
35	3 points	
	subtotal	/ 7
	TOTAL	/ 134

End-of-Year Test - Grade 2

Basic Addition and Subtraction Facts

In problems 1 and 2, your teacher will read you the addition and subtraction questions. Try to answer them as quickly as possible. In each question, he/she will only wait a little while for you to answer, and if you do not say anything, your teacher will move on to the next problem. So just try your best to answer the questions as quickly as you can.

1. Add.

a.	b.	c.	d.
$6 + 7 = $ _____	$7 + 4 = $ _____	$8 + 8 = $ _____	$9 + 5 = $ _____
$9 + 9 = $ _____	$5 + 8 = $ _____	$6 + 6 = $ _____	$7 + 7 = $ _____
$5 + 6 = $ _____	$3 + 9 = $ _____	$2 + 9 = $ _____	$8 + 6 = $ _____
$8 + 7 = $ _____	$5 + 7 = $ _____	$4 + 8 = $ _____	$8 + 9 = $ _____

2. Subtract.

a.	b.	c.	d.
$12 - 3 = $ _____	$11 - 3 = $ _____	$14 - 5 = $ _____	$13 - 4 = $ _____
$15 - 7 = $ _____	$12 - 8 = $ _____	$12 - 4 = $ _____	$15 - 6 = $ _____
$13 - 6 = $ _____	$14 - 6 = $ _____	$18 - 9 = $ _____	$12 - 6 = $ _____
$11 - 7 = $ _____	$16 - 8 = $ _____	$16 - 7 = $ _____	$14 - 7 = $ _____

3. Fill in the missing numbers. The four problems form a fact family.

a.	b.	c.
$2 + \boxed{} = 11$	___ $+$ ___ $= 17$	___ $+$ ___ $=$ ___
$\boxed{} + 2 = 11$	___ $+$ ___ $= 17$	___ $+$ ___ $=$ ___
$11 - 2 = \boxed{}$	$17 - 8 = $ ___	$12 - $ ___ $= 5$
$11 - \boxed{} = 2$	$17 - $ ___ $= $ ___	___ $- $ ___ $= $ ___

Mental Addition and Subtraction with Two-Digit Numbers and Word Problems

4. What is the double of 35?

5. Mary picked 5 apples and Bill picked 9. The children shared all of their apples evenly. How many did each child get?

6. List the even numbers from 10 to 20.

7. Find the difference between 75 and 90.

8. Tim had saved $16. Then Grandmother gave him $10. Now how much more does he need in order to buy a game for $32?

9. Find the missing numbers.

   a. $82 + \underline{\hspace{1cm}} = 90$ **b.** $13 + \underline{\hspace{1cm}} = 21$ **c.** $90 - \underline{\hspace{1cm}} = 83$

10. Calculate the answer in your mind.

a. $59 + 8 = \underline{\hspace{1cm}}$	**b.** $52 + 40 = \underline{\hspace{1cm}}$	**c.** $76 - 50 = \underline{\hspace{1cm}}$
$62 + 8 = \underline{\hspace{1cm}}$	$45 + 9 = \underline{\hspace{1cm}}$	$54 - 23 = \underline{\hspace{1cm}}$

Three-Digit Numbers

11. Write with numbers.

   a. 6 tens 2 hundreds 7 ones = $\underline{\hspace{1.5cm}}$ **b.** 8 ones 9 hundreds = $\underline{\hspace{1.5cm}}$

12. Skip-count by tens.

   568, 578, $\underline{\hspace{1.5cm}}$, $\underline{\hspace{1.5cm}}$, $\underline{\hspace{1.5cm}}$, $\underline{\hspace{1.5cm}}$, $\underline{\hspace{1.5cm}}$

13. Write the numbers in order from the smallest to the greatest.

a. 417, 714, 447	b. 89, 998, 809

14. Calculate the answer in your mind.

a. 560 + 40 = _____	b. 520 − 20 = _____	c. 362 − 30 = _____
560 + 400 = _____	520 − 200 = _____	362 − 300 = _____

15. Compare the expressions and write < , > or = .

a. $100 - 5 - 3 \boxed{} 98 - 6$ b. $40 + 8 + 200 \boxed{} 20 + 800 + 4$

c. $50 + 120 \boxed{} 125$ d. $\frac{1}{2}$ of $800 \boxed{} 399 + 5$

Regrouping in Addition and Subtraction, including Word Problems

16. Add.

a.
$$\begin{array}{r} 3\ 5 \\ 3\ 6 \\ +\ 1\ 2 \\ \hline \end{array}$$

b.
$$\begin{array}{r} 2\ 2\ 4 \\ +\ 4\ 5\ 8 \\ \hline \end{array}$$

c.
$$\begin{array}{r} 4\ 3\ 8 \\ 1\ 7 \\ +\ 2\ 9\ 3 \\ \hline \end{array}$$

17. Subtract. Check by adding the result and what was subtracted.

a. $\begin{array}{r} 6\ 1 \\ -\ 3\ 7 \\ \hline \end{array}$ + _____	b. $\begin{array}{r} 9\ 7\ 0 \\ -\ 2\ 4\ 8 \\ \hline \end{array}$ + _____

18. Jennifer bought two vacuum cleaners for $152 each. What was the total cost?

19. A box contains 450 disks in all. There are 126 music CDs and the rest are DVDs. How many DVDs are in the box?

20. The distance from Vince's home to his Grandmother's home is 218 kilometres. A round trip from his home to her home and back would be how many kilometres?

21. Jane jogs every day. The track is in the shape of a rectangle. One of its sides is 150 metres and another side is 300 metres.

   a. Mark the distances in the picture.

   b. What is the distance when Jane jogs around the track once?

Clock

22. Write the time with *hours:minutes*, and using "past" or "to".

a.	**b.**	**c.**
_____ : _____	_____ : _____	_____ : _____
_____ past _____	_____	_____

23. How much time passes? Fill in the table.

from	3:00	2:00	1 AM	11 AM	8 PM
to	3:05	2:30	8 AM	1 PM	midnight
amount of time					

Money

24. How much money? Write the amount.

a. $_____

b. $_____

25. Find the change, if you buy a snack for $3.35
    and you pay with $4.

26. Brian bought a banana that cost 85¢. He paid with $1.
    What was his change?

51

27. Name the shapes.

    Shape A: _____

    Shape B: _____

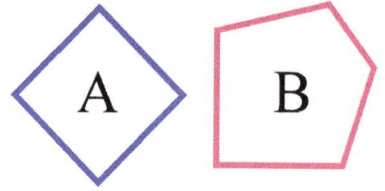

28. **a.** Join the dots in order (A-B-C-D)
    with straight lines. Use a ruler.

    b. What shape is formed?

    _____

    c. Measure the sides of the shape to the nearest centimetre.

    Side AB: about _____   Side BC: about _____

    Side CD: about _____   Side DA: about _____

29. Measure this line to the nearest centimetre.

    ▬▬▬▬▬▬▬▬▬▬▬▬   about _____ cm

30. Which measuring unit or units could you use to find these amounts?
    centimetre (cm), metre (m), or kilometre (km)?

Distance	Unit(s)
how long my pencil is	
the distance from London to Wales	
the height of a wall	
the distance it is to the neighbour's house	

Fractions

31. Divide these shapes. Colour the part you are asked to colour.

a.	b.	c.	d.
Divide this into thirds. Colour $\frac{2}{3}$.	Divide this into halves. Colour $\frac{1}{2}$.	Divide this into halves. Colour $\frac{2}{2}$.	Divide this into fourths. Colour $\frac{3}{4}$.

32. Colour in the given fraction. Compare and write $<$, $>$ or $=$ between the fractions.

a. $\frac{1}{2}$ $\frac{2}{5}$ b. $\frac{4}{6}$ $\frac{3}{4}$ c. $\frac{2}{3}$ $\frac{2}{4}$

The Concept of Multiplication

33. Write a multiplication sentence for each picture.

a. _____ × _____ = _____

b. _____ × _____ = _____

34. Write a multiplication for each addition, and solve.

a. $5 + 5 + 5$	b. $4 + 4 + 4 + 4 + 4$
_____ × _____ = _____	_____ × _____ = _____

35. Solve.

a. $2 \times 5 =$ _____	b. $3 \times 3 =$ _____	c. $3 \times 10 =$ _____

Using the cumulative revisions and the worksheet maker

The student books contain mixed revision lessons which revise concepts from earlier chapters. The curriculum also comes with additional cumulative revision lessons, which are just like the mixed revision lessons in the student books, with a mix of problems covering various topics.

These cumulative revisions are optional; use them as needed. They are named indicating which chapters of the main curriculum the problems in the revision come from. For example, the revision titled "Cumulative Revision, Chapters 1 - 4" includes problems that cover topics from chapters 1-4.

In the digital version of the curriculum, the cumulative revisions are provided both as PDF files and as html files. Again, the html versions are editable.

Both the mixed and cumulative revisions allow you to spot areas that the student has not grasped well or has forgotten. Another sign that the student has not understood a concept or skill is if he/she cannot do word problems in the curriculum that require that concept or skill.

The worksheet maker

The download version of the curriculum includes the worksheet maker as a file, and you can also access the worksheet maker online at

https://www.mathmammoth.com/private/Make_extra_worksheets_grade2.htm

This worksheet maker is the perfect tool to make more problems for students who need more practice. It covers most topics in the curriculum, excluding word problems, and most people find it to be a very helpful addition to the curriculum.

Options for additional review

When you find a topic or concept your student needs more practice with, you have several options:

1. Check if the worksheet maker lets you make worksheets for that topic (for example, conversions between measuring units or equivalent fractions).

2. Check for any online games and resources in the Introduction part of the particular chapter in which this topic or concept was taught.

3. If you have the digital version, you could simply reprint the lesson from the student worktext, and have the student restudy that.

4. Perhaps you only assigned 1/2 or 2/3 of the exercise sets in the student book at first, and can now use the remaining exercises.

5. Check if our online practice area at https://www.mathmammoth.com/practice/ has something for that topic. We are constantly adding more exercises and games to this.

6. Khan Academy has free online exercises, articles, and videos for most any maths topic imaginable.

Cumulative Revision, Grade 2, Chapters 1 - 2

1. Solve the problems. Fill in the doubles chart. **It has a pattern!** Find it!

a. It will take Andrew 16 hours to clean the park. He cleaned half of it yesterday. How many hours will he still have to work?

$10 + 10 =$ _____

$15 + 15 =$ _____

$20 + 20 =$ _____

b. What is double 12?

$25 + 25 =$ _____

c. Ashley and Erika divided $30 evenly. Then Erika bought a gift for $6. How much money does Erika have now?

$30 + 30 =$ _____

$35 + 35 =$ _____

$40 + 40 =$ _____

d. Terrance has saved $20. That is just half of what he needs to buy a train set. How much does the train set cost?

2. Add and subtract whole tens.

a.

$+30$ $+10$ -20 -20

10

b.

-40 $+30$ -40 $+20$

60

3. Write the time using *hours:minutes*.

a.

_____ : _____

b.

_____ : _____

c.

_____ : _____

d.

_____ : _____

4. **a.** Beth began watching a film about sea animals at 20 to 4. She stopped watching it at 15 past 4. Write those two times in the *hours:minutes* way.

_____ : _____ and _____ : _____

b. Jason began walking his dog at 11 AM and stopped at noon. How long did he walk his dog?

c. Brad's rooster crowed for half an hour, starting at 5 AM. At what time did it stop crowing?

5. Fill in the missing numbers. The four problems form a fact family.

a. $3 + \boxed{} = 9$

$\boxed{} + 3 = 9$

$9 - 3 = \boxed{}$

$9 - \boxed{} = 3$

b. _____ + _____ = 10

_____ + _____ = 10

$10 - 4 = \boxed{}$

$10 - \boxed{} = 4$

c. _____ + _____ = _____

_____ + _____ = _____

$8 -$ _____ $= 3$

_____ $-$ _____ $=$ _____

6. Find the letters, and find out what Bruce got for his birthday.

The second row from the top,
the second letter from the left.   ——

The fourth row from the top,
the fifth letter from the left.   ——

The first row from the top,
the fifth letter from the right.   ——

The fifth row from the bottom,
the second letter from the right.   ——

The 1st row from the bottom,
the 1st letter from the left.   ——

The sixth row from the top,
the third letter from the right.   ——

The 3rd row from the top,
the 2nd letter from the left.   ——

E	S	H	A	B	G	P
B	A	E	N	I	V	S
W	E	K	P	T	O	F
J	D	A	U	-	W	M
Y	K	Z	N	Y	I	C
U	D	T	S	S	Q	R
R	T	H	A	V	E	L

Cumulative Revision, Grade 2, Chapters 1 - 3

1. Add and find the missing numbers (addends).

a.	b.	c.	d.
$6 + 7 =$ _____	$9 + 7 =$ _____	$5 +$ _____ $= 14$	$8 +$ _____ $= 15$
$8 + 9 =$ _____	$5 + 8 =$ _____	$8 +$ _____ $= 16$	$7 +$ _____ $= 14$

2. How many hours is it?

from	9 AM	6 AM	11 AM	12 AM	10 AM
to	1 PM	8 PM	4 PM	12 PM	2 PM
hours					

3. **a.** How many Tuesdays are there in January?
   (See the calendar on the right.)

   b. Jane visits her parents every third Sunday of the
   month. What day will she visit them in January?

January

Su	Mo	Tu	We	Th	Fr	Sa
		1	2	3	4	5
6	7	8	9	10	11	12
13	14	15	16	17	18	19
20	21	22	23	24	25	26
27	28	29	30	31		

4. Solve.

   a. Jenny practises playing the piano for 2 hours. She stopped
   practising at 2 PM. What time did she *start* practising?

   b. Grandmother sleeps 1/4 of the day's hours. (One day has 24 hours.)
   How many hours does Grandmother sleep each day?

5. Add and subtract whole tens.

a. $77 + 20 =$ _____	**b.** $18 + 50 =$ _____	**c.** $54 + 40 =$ _____
$64 - 30 =$ _____	$43 - 20 =$ _____	$98 - 60 =$ _____

6. Add more. Find the difference.

a. $18 + \underline{\hspace{1.5cm}} = 22$	**b.** $75 + \underline{\hspace{1.5cm}} = 80$	**c.** $56 + \underline{\hspace{1.5cm}} = 59$
d. The difference between 8 and 12 is _____.	**e.** The difference between 43 and 49 is _____.	**f.** The difference between 21 and 30 is _____.

7. Subtract. Think about the difference.

a. $85 - 80 = \underline{\hspace{1cm}}$ $46 - 42 = \underline{\hspace{1cm}}$	**b.** $76 - 71 = \underline{\hspace{1cm}}$ $99 - 89 = \underline{\hspace{1cm}}$	**c.** $20 - 17 = \underline{\hspace{1cm}}$ $70 - 67 = \underline{\hspace{1cm}}$

8. For each addition, write a matching subtraction (using the same numbers).

a. $8 + \boxed{} = 14$ $\underline{\hspace{1cm}} - \underline{\hspace{1cm}} = \boxed{}$	**b.** $5 + \boxed{} = 14$ $\underline{\hspace{1cm}} - \underline{\hspace{1cm}} = \boxed{}$	**c.** $6 + \boxed{} = 12$ $\underline{\hspace{1cm}} - \underline{\hspace{1cm}} = \boxed{}$

9. Subtract.

a. $12 - 7 = \underline{\hspace{1cm}}$ $17 - 9 = \underline{\hspace{1cm}}$ $11 - 8 = \underline{\hspace{1cm}}$	**b.** $14 - 8 = \underline{\hspace{1cm}}$ $12 - 8 = \underline{\hspace{1cm}}$ $13 - 7 = \underline{\hspace{1cm}}$	**c.** $11 - 6 = \underline{\hspace{1cm}}$ $13 - 8 = \underline{\hspace{1cm}}$ $16 - 9 = \underline{\hspace{1cm}}$	**d.** $15 - 7 = \underline{\hspace{1cm}}$ $14 - 9 = \underline{\hspace{1cm}}$ $15 - 9 = \underline{\hspace{1cm}}$

10. Detective Bryant is a math sleuth. He was out to get the fact family. He had found number 13, but two numbers were missing. Help him find the fact family!

He found a clue under the sofa: "Look in the cookie jar!" In the cookie jar there were half a dozen cookies left. Bryant said, "That is one of my missing numbers!"

Can you figure out the other missing number now? Then, write the fact family.

$\underline{\hspace{1cm}} + \underline{\hspace{1cm}} = \underline{\hspace{1.5cm}}$ $\underline{\hspace{1.5cm}} - \underline{\hspace{1cm}} = \underline{\hspace{1cm}}$

$\underline{\hspace{1cm}} + \underline{\hspace{1cm}} = \underline{\hspace{1.5cm}}$ $\underline{\hspace{1.5cm}} - \underline{\hspace{1cm}} = \underline{\hspace{1cm}}$

The case is solved!

Cumulative Revision, Grade 2, Chapters 1 - 4

1. Solve the problems.

a. One-half of the boys in the class are studying math.
   The other seven boys are reading.
   How many boys are in the class?

b. Mark has $8. Andrew has double that much money.
   How much money does Andrew have?

   How much money do the two boys have together?

2. Add mentally.

a. $28 + 4 =$ _____	**b.** $39 + 9 =$ _____	**c.** $44 + 5 + 4 =$ _____
$28 + 40 =$ _____	$30 + 29 =$ _____	$7 + 8 + 9 + 4 =$ _____

3. A few years ago a small camera cost $67.
   Now it has doubled in price.
   How much does it cost now?

$+$

4. Write the time using "past", "to", "half past", or "o'clock".

a. 7:25 *25 past 7*	**b.** 5:10 _____
c. 5:50 _____	**d.** 12:40 _____
e. 12:30 _____	**f.** 11:00 _____

5. Write the numbers so that ones and tens are in their own columns. Add.

a. $44 + 37$ **b.** $9 + 26$ **c.** $26 + 8 + 47$ **d.** $25 + 57 + 38$

6. Seth made six cards for his party. He put the cards at the plate of each guest. It took Seth 10 minutes to make one card.

   a. How long did it take Seth to make all 6 cards?

   b. If he started making the cards at noon, at what time did he finish his project?

7. Fill in the missing numbers.

a. $24 + 8 = \bigcirc$	**b.** $16 - 7 = \bigcirc$	**c.** $17 - 9 = \bigcirc$
d. $\bigcirc - 6 = 5$	**e.** $\bigcirc - 20 = 7$	**f.** $\bigcirc - 5 = 31$

8. You bought three towels for $18 each and a vase for $25. Find the total cost.

Cumulative Revision, Grade 2, Chapters 1 - 5

1. Write two additions and two subtractions for each picture. The box with a "T" is a ten.

a. [T / T] and [T T / T T]

_____ + _____ = _____

_____ + _____ = _____

_____ − _____ = _____

_____ − _____ = _____

b. [T T / ● ● ● ●] and [dots]

_____ + _____ = _____

_____ + _____ = _____

_____ − _____ = _____

_____ − _____ = _____

2. Add.

    a. 29 + 90   **b.** 93 + 46   **c.** 24 + 35 + 48   **d.** 22 + 47 + 9

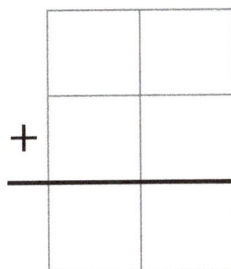
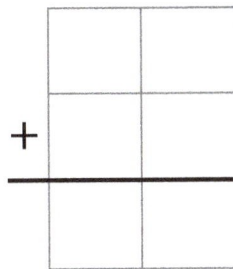
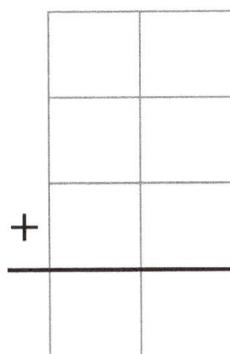
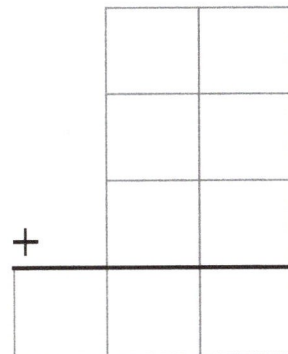

3. Solve.

a. Choose the letters from the given word to make a new word.

    M A M M A L

   ____   ____   ____   ____   ____

   6th   6th   5th   3rd   2nd

b. Put the letters in the given order to make a new word.

    L   O   I   D   R   N   A

    7th   1st   4th   3rd   2nd   5th   6th

  O   ____   ____   ____   ____   ____   ____

4. Write the time that the clock shows, and the time 5 minutes later.

	a. _____ : _____	b. _____ : _____	c. _____ : _____	d. _____ : _____
5 min. later →	_____ : _____	_____ : _____	_____ : _____	_____ : _____

5. A flower vase has 15 flowers. Some are red, some are blue, and some are yellow. Four of the flowers are red and five are yellow. How many are blue?

6. Write < , > or = . You can often compare without calculating!

a. $8 + 8$ ☐ $9 + 8$ **b.** $30 - 8$ ☐ $30 - 9$ **c.** $\frac{1}{2}$ of 16 ☐ 16

d. $35 + 7$ ☐ $35 + 8$ **e.** $40 - 6$ ☐ $40 - 9$ **f.** $14 - 7$ ☐ $16 - 8$

7. Add by adding tens and ones separately.

a. $\quad$ $36 + 22$	**b.** $\quad$ $72 + 18$
$30 + 20 + 6 + 2$	$70 + 10 + 2 + 8$
_____ + _____ = _____	_____ + _____ = _____
c. $\quad$ $54 + 37$	**d.** $\quad$ $24 + 55$
_____ + _____ = _____	_____ + _____ = _____

8. Count in 10s and 50s, and fill in the grids.

a.	464	474						

b.	400	450						

Cumulative Revision, Grade 2, Chapters 1 - 6

1. Write what place the teddy bear is in using ordinal numbers.

   a. The _____ place from the left.

   b. The _____ place from the right.

   c. The _____ place from the left.

   d. The _____ place from the right.

2. Joel is on the track team. He spends a half-hour for warm-up
   exercises, an hour running, and thirty minutes jumping hurdles.
   How much time does he spend practising?

3. One book costs $12 and another costs $5 more than that.
   If you buy both, what is the total cost?

4. Add.

   a.
   ```
     3 7
     1 8
   + 4 3
   ─────
   ```

   b.
   ```
     2 9
     8 0
   + 3 6
   ─────
   ```

   c.
   ```
     5 4
     1 3
   + 7 6
   ─────
   ```

   d.
   ```
     3 6
       9
   + 4 3
   ─────
   ```

   e.
   ```
     2 8
     1 8
   + 3 6
   ─────
   ```

5. Complete the **next whole ten**.

a.	b.	c.
$66 + \underline{\quad} = 70$	$31 + 3 + \underline{\quad} = 40$	$47 + \underline{\quad} + 1 = 50$
$92 + \underline{\quad} = \underline{\quad}$	$63 + 2 + \underline{\quad} = 70$	$32 + \underline{\quad} + 2 = 40$

6. Write the time with *hours:minutes*, and using "past" or "to".

a.	b.	c.	d.
_____ : _____	_____ : _____	_____ : _____	_____ : _____
_____ past _____	_____	_____	_____

7. Draw six dots randomly and join them like a dot-to-dot. Use a ruler. What shape do you get?
   (Hint: It will not be a straight line.)

8. Colour <u>one whole pie</u>. Write <u>one</u> as a fraction, in many different ways.

a. $1 = \dfrac{\quad}{\quad}$ b. $1 = \dfrac{\quad}{\quad}$ c. $1 = \dfrac{\quad}{\quad}$

9. Divide these shapes. Then colour as you are asked to.

a.	b.	c.	d.
Divide this into halves. Colour $\dfrac{1}{2}$.	Divide this into fourths. Colour $\dfrac{3}{4}$.	Divide this into thirds. Colour $\dfrac{1}{3}$.	Divide this into fourths. Colour $\dfrac{1}{4}$

Cumulative Revision, Grade 2, Chapters 1 - 7

1. Solve.

> **a.** Penny needs 8 apples to make one pie, and she wants to make two pies for a bake sale. Penny already has 10 apples. How many more apples does Penny need to buy?

> **b.** Jesse and Blake are making pencil holders. They sold one for $18. How much would two pencil holders cost?

> **c.** Ramona has four cats. One of them had kittens. Now she has double as many cats as before.
>
> How many cats does Ramona have now?
>
> How many of them are kittens?

2. Draw the hands on the clock faces to show the given time.

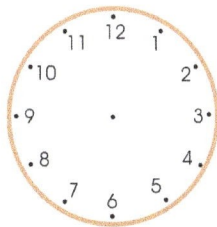 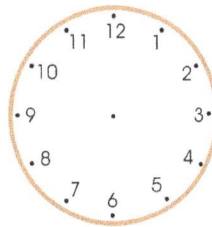 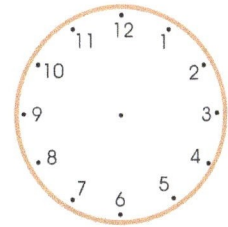

         a. 1:25   **b.** 3:15   **c.** 15 to 3

3. Find the pattern and continue it.

a. $577 - 10 = $ _____	**b.** $926 - 0 = $ _____
$577 - 20 = $ _____	$926 - 100 = $ _____
$577 - 30 = $ _____	$926 - 200 = $ _____
$577 - $ ___ $ = $ _____	___ $ - $ ___ $ = $ _____
___ $ - $ ___ $ = $ _____	___ $ - $ ___ $ = $ _____
___ $ - $ ___ $ = $ _____	___ $ - $ ___ $ = $ _____

4. Solve.

a. $9 + 8 =$ _____	b. $8 + 8 =$ _____	c. $5 + 8 =$ _____
$5 + 6 =$ _____	$6 + 6 =$ _____	$4 + 7 =$ _____
$7 + 7 =$ _____	$9 + 7 =$ _____	$6 + 8 =$ _____

5. Anna drew some shapes in her notebook.
   She needs you to help her label them.

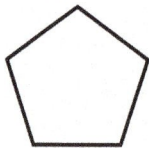

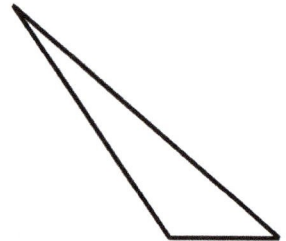

a._____   b. _____   c. _____

6. Draw in the grid a rectangle that is
   6 units wide and 3 units long.

   How many squares are inside it? _____ squares

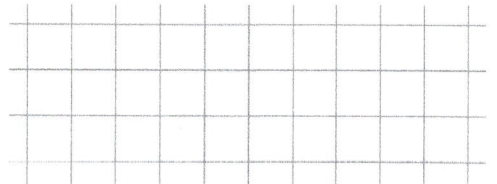

7. Fill in the missing numbers and words in the charts below.

Ordinal Number	Name	Ordinal Number	Name
	first	8th	
2nd			ninth
		10th	
	fourth		
			thirteenth
	seventh		

Cumulative Revision, Grade 2, Chapters 1 - 8

1. Cross out and subtract. Subtract also in columns!

	c. 62 – 25	d. 83 – 46
T T T (dots) **a.** 40 – 29 = _____ T T T T (dots) **b.** 50 – 28 = _____		

2. Add.

a.	b.	c.	d.
6 + 3 = _____	7 + 5 = _____	8 + 6 = _____	9 + 9 = _____
6 + 10 = _____	7 + 8 = _____	8 + 7 = _____	9 + 4 = _____

3. Find these differences. Think of adding more.

a. 17 – 11 = _____ Think: 11 + ____ = 17	**b.** 43 – 37 = _____ Think: 37 + ____ = 43	**c.** 66 – 59 = _____ Think: 59 + ____ = 66
d. 35 – 28 = _____	**e.** 80 – 77 = _____	**f.** 100 – 94 = _____

4. Find what was subtracted.

– ☐   – ☐   – ☐   – ☐   – ☐   – ☐   – ☐

79   75   71   68   65   59   57   52

5. Write the times using hours : minutes.

a. 15 past 6	b. 20 to 3	c. 5 past 10	d. half past 3
_____ : _____	_____ : _____	_____ : _____	_____ : _____
e. 15 to 8	f. 20 to 12	g. 5 to 1	h. 25 past 1
_____ : _____	_____ : _____	_____ : _____	_____ : _____

6. Daniel weighs 62 kilograms. His sister weighs twenty kilograms less than he does.

   a. How much does Daniel's sister weigh?

   b. How much do they weigh together?

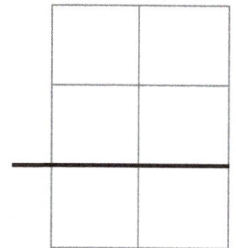

7. a. Draw a rectangle. Use a ruler to make it as neat as you can!

   b. Draw a line through the rectangle from one corner to the opposite corner.

   c. What shapes are formed now?

8. a. In each picture, colour TWO slices of the whole pie, and write the fraction.

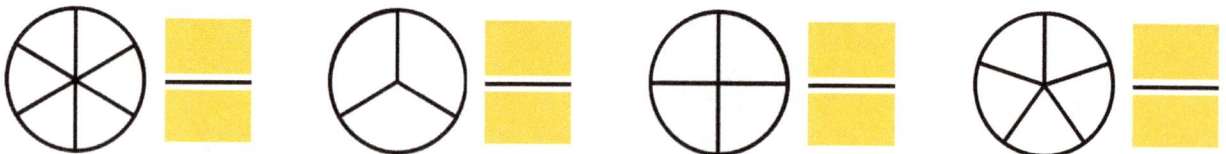

   b. Now, find the largest fraction (the one that has the most to eat).

9. Compare, and write < or > .

   a. 106 ☐ 120   b. 141 ☐ 114   c. 700 + 80 + 9 ☐ 90 + 8 + 700

Cumulative Revision, Grade 2, Chapters 1 - 9

1. Colour the part indicated.

a. $\dfrac{1}{2}$ b. $\dfrac{1}{4}$ c. $\dfrac{3}{4}$ d. $\dfrac{2}{4}$ e. $\dfrac{4}{4}$ f. $\dfrac{2}{2}$

2. How many hours pass?

a. From 3:00 to 8:00 _____ hours	**d.** From noon to 4 PM _____ hours
b. From 6 AM to 1 PM _____ hours	**e.** From 7 PM to 11 PM _____ hours
c. From 7 PM to midnight _____ hours	**f.** From 10 AM to 2 PM _____ hours

3. Add and subtract.

a. $7 + 8 =$ _____	b. $8 +$ _____ $= 13$	c. $14 -$ _____ $= 7$
$4 + 9 =$ _____	$8 -$ _____ $= 2$	$6 +$ _____ $= 11$
$15 - 9 =$ _____	$4 +$ _____ $= 11$	$19 -$ _____ $= 12$

4. Add money amounts in columns.

a. $\$1.35 + \$1.55$ b. $\$2.25 + \$1.90$ c. $\$5.55 + \$2.70$

5. **a.** How many corners does the shape have? _____

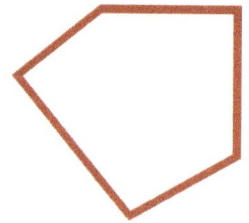

   b. What is the shape called? _____

6. In a game, Amanda has 15 marbles and John has 5 fewer marbles than Amanda.
   How many does John have?

   How many marbles do Amanda and John have together?

7. Jeremy ate 4 slices of pie, which was two fewer pieces than what Emma ate.
   How many did Emma eat?

8. Add and subtract mentally.

a.	b.	c.
507 + 30 = _____	640 − 40 = _____	552 − 20 = _____
507 + 300 = _____	640 − 400 = _____	552 − 200 = _____

9. Subtract. Check by adding!

a.
Check:
```
    6 7 0
  − 3 3 8    + _____
  _____
```

b.
Check:
```
    5 4 1
  − 2 7 1    + _____
  _____
```

10. Count up to find the change. You can draw in the coins to help you.

a. $3.80
Customer gives $5

Change: _____

b. $6.85
Customer gives $10

Change: _____

Cumulative Revision, Grade 2, Chapters 1 - 10

1. Colour in the chart all the even numbers.

1	2	3	4	5	6	7	8	9	10
11	12	13	14	15	16	17	18	19	20
21	22	23	24	25	26	27	28	29	30

2. **a.** Today is 5 January. I am going away for three weeks and two days. What day will I return? (See the calendar on the right.)

January

Su	Mo	Tu	We	Th	Fr	Sa
		1	2	3	4	5
6	7	8	9	10	11	12
13	14	15	16	17	18	19
20	21	22	23	24	25	26
27	28	29	30	31		

   b. I went to the gym every Wednesday in January. What were the dates I went to the gym?

3. Aunt Sarah gave Larry and Patty $30. The children shared the money equally. Patty already had $5 in her piggy bank. How much money does Patty have now?

4. Subtract.

a.
$$975 - 246$$

b.
$$629 - 189$$

c.
$$514 - 323$$

d.
$$650 - 126$$

5. Find the missing numbers.

a. $900 + \boxed{} = 914$	**b.** $620 + \boxed{} = 680$	**c.** $600 - \boxed{} = 570$
d. $\boxed{} - 20 = 40$	**e.** $\boxed{} - 70 = 70$	**f.** $572 - \boxed{} = 512$

6. Write the amounts using the $ sign and a decimal point.

a.	b.	c.
$_____	$_____	$_____

7. Molly bought an eraser for $2.95 and gave the shopkeeper $5.
   What was her change?

8. What is the total if you have six dimes,
   four nickels and a quarter?

9. Weigh yourself. I weigh _____. Now weigh yourself holding as many

   books as you can hold. I weigh _____ with the books.

   How much do the books weigh? _____

10. Solve.

a. Martha has saved $25. She wants to buy a sewing kit for $45. After she earns $15, can she buy it?	b. Find the cost of buying three plants for $17 each.	c. One sack of potatoes weighs 22 kg. How much would four sacks weigh?
+	+	+

www.ingramcontent.com/pod-product-compliance
Lightning Source LLC
Chambersburg PA
CBHW051352200326
41521CB00014B/2547